presented

narrative of

the course

the

48

first proclamation of the

But the God who has thus [illegible]

[illegible] course of direction [illegible]

[illegible] longer dwells [illegible]

Olympia

presented

the courts

the ter

... the first proclaiming of the ...

But the God who has thus fore-

... its coming destruction ...

... no longer dwells ...

presented

narrative of

the course

the let

and the first proclaiming of the

But the God who has thus retur

and His coming destruction (that

no longer dwells in

Olympia

www.ingramcontent.com/pod-product-compliance
Lightning Source LLC
LaVergne TN
LVHW080558200726
843510LV00004B/949

* 9 7 8 3 3 4 7 0 2 5 7 0 7 *